This report is part of a series of reports on technical rescue incidents across the United States. Technical rescue has become increasingly recognized as an important element in integrated emergency response. Technical rescue generally includes the following rescue disciplines: confined space rescue, rope rescue, trench/collapse rescue, ice/water rescue, and agricultural and industrial rescue. The intent of these reports is to share information about recent technical rescue incidents with rescuers across the country. The investigation reports, such as this one, provide detailed information about the magnitude and nature of the incidents themselves; how the response to the incidents was carried out and managed; the impact of these incidents on emergency responders and the emergency response systems in the community; and the lessons learned. The U.S. Fire Administration greatly appreciates the cooperation and information it has received from the tire service, county and state officials, and other emergency responders while preparing these reports.

This report was produced under contract EMW-94-C-4436. Any opinions, findings, conclusions, or recommendations expressed in this publication do not necessarily reflect the views of the U.S. Fire Administration or the Federal Emergency Management Agency.

Additional copies of this report can be ordered from the U.S. Fire Administration, 16825 South Seton Avenue, Emmitsburg, MD 21727.

Search and Rescue Operations in Georgia
During Major Floods
July 1994

Local Contacts:

Don Etheridge,
  Supervisor of Fire Academy
Georgia Fire Academy
Georgia Public Safety Training Center
1000 Indian Springs Drive
Forsyth, GA 31029-9599
(912) 993-4670

David White,
  Deputy Chief
Spalding County Fire Department
Griffin, GA
(404) 228-2129

Chief Hinson
  Deputy Chief Hartley
Macon Fire Department
Macon, GA
(912) 751-5180

Chief Henry Fields
Albany Fire Department
Albany, GA
(912) 431-3262

Chief Stephen R.M. Moreno
Americus Fire Department
Responsible for SAR under the direction
of Sheriff Randy Howard, Director of Sumter
County Emergency Management Agency
Americus, GA
(912) 924-3213

Alan Greene,
 Director
Emergency Management Agency
Jones County, GA
(912) 986-6672

OVERVIEW

During a three-week span in July 1994, the state of Georgia sustained its worst damage ever as the result of flooding. In addition to billions of dollars of property loss, more than 50 lives were lost, and thousands of homeowners were displaced for several weeks. Municipal services in several communities were severely disrupted, as riverscut communities into several pieces and water supplies were contaminated. The flood was the direct result of the stalled Hurricane Alberto, which was later downgraded to a tropical storm. The storm stalled on-shore for more than a week, with its low center hanging over the middle of the state, between Albany and Macon.

Flood disasters have occurred with increasing frequency throughout the United States in the last decade of the 20th century, mostly due to man's growing encroachment on the natural flood plains of North American rivers. The floods along the, length of the Mississippi in 1993, and the flooding in California in 1992 and 1994-1995 can be directly attributed to the failure of such narrowing and dam-building efforts.

In Georgia the drainages affected had no major dams. The primary cause of the flooding was a tropical low pressure area that stalled over central Georgia. In the course of a seven-day period more than 40 inches of rain fell. In some areas the last 24 inches fell in 24 hours.

The primary drainages affected were along the Ocmulgee and Flint Rivers in central Georgia, though localized flooding caused profound problems elsewhere as well. While local authorities noted the steady heavy rains, there was little local warning of flooding, which increased in severity and occurrence generally from north to south. Due to this increase in the volume of water which moved south along the river drainages, search and rescue operations needed to be substantially increased.

I. LOCAL SEARCH AND RESCUE OPERATIONS

Griffin and Spalding County

The Spalding County Fire Department has 50 paid personnel and 12 volunteers staffing 5 stations. On July 5, 1994 Chief David White noted that Atlanta weathercasters were reporting localized flooding along the Chatahoochy River and Nancy Creek (events that usually took at least two weeks of steady rain.) Sensing that problems were about to occur, Chief White began preplanning for recovery efforts along the Spalding County waterways.

Since the Flint River cuts the district in half, Chief White stationed extra firefighters and equipment on the west side of the river. But since Station Three was only 150 yards from the river he ordered the duty crew to closely watch the river. He also planned potential evacuation of the station. Then Chief White contacted nearby Fayette County, so he could house his personnel in their station at Brooks, Georgia. He also sent a private ambulance to the west side of the river, and ordered personnel to secure the LP gas tanks at the station.

By the morning of July 6th, Station Three was submerged in four feet of water and had to be abandoned.

The next 48 hours were the most critical for Spalding County. Police and fire units made approximately 30 rescues. The county also suffered two fatalities. The flooding caused 57 bridges and a number of roads to wash out. Ten of the bridges still remained impassable six months later.

Incidents included two vehicles being washed away in a low-water crossing; four residents trapped in a house; animal rescues; and a mutual-aid request from nearby Henry County's Dive Team to search for two teenagers in a cheap raft on the Tuolugu River. The teenagers were later found dead in a debris pile.

Rescue workers, who had no previous training in river rescue, found themselves getting on-the-job training. After discovering that only a few feet of water could move a car, and that fast-flowing water would sweep someone wading in water only hip-deep, firefighters rigged ropes to help rescue 18 stranded residents across flooded Wildcat Creek.

Using a flood plain map from the State of Georgia which showed the projected flood plains at various levels, rescuers were able to identify and predict problem areas. However, many local residents refused to stay out of the sewage contaminated water. Officials finally had to threaten arrest to keep the residents safe.

The City of Griffin utilized the County convict crews and road department equipment for levees and sandbagging operations. Helicopters from the Department of Natural Resources, Decalb and Clayton Counties were used for search and surveillance. Additionally, the Department of Natural Resources sent five boats to assist local agencies.

The water levels rose and receded so quickly in the Griffin area that the local Emergency Operations Center was not activated, which meant that Georgia Emergency Management Agency aid was not utilized.

Macon Bibb Fire Department - Chiefs Hinson and J.P. Hartley

With more than 120,000 people the City of Macon and environs sit on the Ohomagee River, alongside Interstate 75, the main north-south freeway in Georgia, and 60 miles further south from Griffin.

The Ohomagee River also rose and receded quickly, although the aftereffects were more profound. Experts rated this rise of the Ohomagee as a 500-year flood. Local officials knew they were going to have a flood, however no one realized it was going to be so destructive.

The initial flooding on July 6 forced the evacuation of a Best Western Motel and a Days Inn. It also forced the closure of I-75, which resulted in traffic being backed up for miles in both directions. Several housing projects were cut off by this closure, and the City of Macon's Emergency Operations Center was activated. The city also requested assistance from Georgia Emergency Management Agency (GEMA), and, on declaration of the federal disaster, the Federal Emergency Management Agency (FEMA) as well.

Potentially, the most profound problem was the flooding of the city's water treatment plant, located in the high water channel of the river. Though planning had been underway to move it for several years, it was still located in this exposed spot. The flooding of the plant cut off drinking water to the entire population, and eliminated the city's ability to treat sewage.

For the next 21 days Macon residents were without fresh water. Residents had to drink water brought in by contracted water tankers to several distribution points. The city also requested 1300 portable toilets, which were distributed and serviced by other contractors. All of these facilities were ordered through GEMA or FEMA. Operations to distribute the water, monitor disease control, and manage the field toilets required the majority of the fire department's time during the next three weeks. However, due to the quick "up and down" nature of the flooding, after the first two days it was not necessary to call in off-duty personnel.

Fortunately, during this time there were no major fires, and only one haz-mat incident occurred, which involved an overturned diesel tanker truck. The fire department was able to keep the hydrant system intact, maintaining some two million gallons of fresh water in above-ground reservoirs as a last-ditch back-up for firefighting and consumption.

Officials indicated that GEMA and FEMA acted quickly on the city's behalf. They sent in National Guard military police to help in traffic control, 134 officers and some boats from the Georgia Fishery Commission, and the already mentioned fleet of tanker trucks, and portable toilets. Helicopters were used to patrol for possible levee failures, particularly south of the city. During the emergency period tanker trucks were also used to transport water to the local telephone facilities, which required large amounts of fresh water to maintain their fiber-optics operations.

During the height of the emergency, the fire department moved its operations out of the city E.O.C. Since elected officials do not often participate in disaster drills, the noise and confusion of the E.O.C. was eliminated by this relocation. While this move did lead to some duplication of effort and orders for materials, resulting emergencies were handled more effectively. The telephone company supplied cellular phones and service to the emergency services-which was a tremendous aid in communications.

In Macon, tactical operations were declared complete when the fresh water plant finally restored its service. New equipment was brought in by railroad flatcars, including new motors and filters. Three hundred thousand gallons of fresh water had to be trucked in to clean *each* of 22 filters-a major task. While local residents did not have any idea the water would rise as high as it did, plans to relocate the water facility were not accelerated.

"Rescues" for the Macon area consisted only of evacuation of low lying structures. There were very few stuck vehicles and none of those were trapped in fast moving water.

The only exciting incident involved a "self-dispatched" person who had a Hovercraft available to him. While demonstrating it to two firefighters on board, it turned over, putting all three into the water only a few miles upstream of a bridge under which there was virtually no clearance. All three made it successfully under the bridge and were rescued downstream. However, the Hovercraft was a total loss.

In Albany, Fire Chief and EMA Director Henry Fields had plenty of warning that the water was on its way, and were aware of the problems occurring upstream along the Flint River, which divides Albany. Initial warnings were that the river would rise to 28 feet - seven feet over flood stage. Accordingly, local officials prepared for a flooding potential of 37 feet. During this event the water actually rose to 44 feet.

Department heads were notified early on July 5th that the EOC was going to be activated later that evening. Following pre-plans, flyers were distributed to areas that were going to be potentially inundated, the media was informed to start making public service announcements, and off-duty personnel were notified.

At 6:00 p.m. the EOC was activated in a classroom at the fire station. At 1:30 p.m. on July 6th, mutual aid was requested from nearby Lee County for two boys trapped in a tree by rising water. No aid was sent because almost simultaneously, Albany's Lovers Home Road started to flood. This resulted in the high school and an immediate care center being ordered to evacuate.

7

The EOC quickly became an inadequate working area. With all department heads talking independently to GEMA, duplicate ordering of resources started to occur. Further confusion ensued as all materials were delivered to the fire station. Because the city and county were on different radio systems, dispatching confusion occurred. The EOC phone system was quickly jammed and cellular phones with priority over-rides were ordered. While standard report forms were in the EOC, the volume of calls resulted in the forms not being used uniformly by all departments.

The flooding of the Flint River cut the City of Albany in half and street flooding created further islands. There were five fatalities, none due to lack of notification, but to refusal to leave when ordered. Several vehicles drove around barricades and were also trapped. Local officials had no enforcement powers and were unable to cite the drivers and residents. Law enforcement made 15 to 20 arrests for looting. Several people were also arrested for violating an enforced 6:00 p.m. to 8:00 a.m. curfew. National Guard and police units patrolled using boats and HumVees.

In Albany, Chief Fields was able to call on 170 sworn personnel in 10 stations, and nearly 60 EMA volunteers. Local and mutual aid personnel made approximately 4,000 rescues and assists.

One early and substantial problem involved the inundation of local cemeteries. Several hundred caskets in various stages of deterioration started popping out of the ground. Disintegrating caskets put their remains into the river. Using information from similar situations during the Missouri flooding, officials from the coroner's office and the Georgia Bureau of Investigation organized a field morgue, bringing in refrigeration trucks to hold the remains for identification. Due to this situation and general contamination, all personnel received tetanus shots early in the operations. As of December 1994, several dozen sets of remains were still not identified.

By Friday, July 7, a Presidential Disaster Declaration had been made, and by Sunday July 9, FEMA had set up three disaster assistance offices. While requested support from GEMA arrived timely, FEMA's help with federal assets was generally slower.

Available flood plain maps quickly became useless as the water continued to rise. Since both bridges across the Flint River were under water, getting from one side to the other resulted in a 100-mile drive. A separate command post for the east side of the city was set up and officials and workers were shuttled back and forth using State of Georgia helicopters.

Operations remained at a steady level for several days until flood waters started to recede. Two nursing homes were evacuated. Levees were reinforced near the local community college on the east side of the river. Twenty-eight shelters were opened for evacuees in schools and churches. Portable pumps to empty buildings were ordered, received, and used. The two "john" boats used by the fire department proved inadequate, both because of their small size and the repeated loss of propellers in shallow waters. Propane tanks kept breaking loose, and utility repairs were a constant problem. The city water supply, however, was never contaminated.

Actions after the flooding included imposing a local 1% sales tax to create a new radio system, and expand and modernize the EOC. Instead of a critique of operations, a four-hour class on the Incident Command System was held for all elected officials. Since the last major flood in 1979 only covered half the affected area, new flood plain maps will reflect a potential 50-foot rise. As it was, the affected area stretched two miles from the river in places, with standing water four to five feet deep.

Finally, local discussions of property condemnations affecting 900 residences and 1500 other structures in flood-prone areas were being conducted in December 1994. Such condemned property areas provide land for floods to inundate without further property loss.

Americus and Sumter Counties were perhaps the hardest hit areas despite being located about 15 miles west of the Flint River. The area has three main drainages that dump into the Muhalee River, which in turn meets the Flint near Albany. There are about 11 significant earthen dams in the immediate area, most holding water for farming and irrigation purposes.

Fire Chief Moreno assists the director of the Sumter County Emergency Management Agency, Sheriff Randy Howard. The tire department in Americus provides contract services to a large area around Americus and is staffed with a total of 43 personnel in two stations.

On July 5, the rain reached a crescendo, inundating the area with 24 inches of rainfall in just 24 hours. Earthen dams started to collapse, causing flash floods that traveled down the drainages and across roads. There was no time or warning to deploy barricades, stop traffic, or evacuate potential problem sites.

There were 15 fatalities. Of these, 13 victims washed away in their cars, and two more victims were trapped in a house on the uphill side of a railroad grade. (The grade that ended up acting as a dam.)

A portion of the problem was caused by dam failures sending walls of water downstream to the next darn, and so forth-causing increasing devastation downstream.

Early in operations firefighters almost lost an engine to rising waters. Radios and phones ceased working at 2:00 a.m. on July 6 for several hours.

Despite these problems, rescuers were able to make 43 successful rescues, mostly using ropes to reach vehicles and structures. This despite the lack of any specific training in swift water rescue. The fire department received 70 calls for assistance between 11:00 p.m. and 6:30 a.m. on July 6. In order to make sure that all personnel were wearing lifejackets, emergency purchase orders were taken to local sporting goods stores.

Unlike previously mentioned areas, there were at least three working fires in the area, including a major factory fire in the nearby town of Ogelthorpe that firefighters could not reach due to surrounding water. Health dangers for rescuers were created by hazardous materials, including raw sewage and several propane tanks which had been torn loose. As a preventative measure, firefighters received tetanus boosters and began a hepatitis series. Additionally, stringent decontamination procedures were implemented for returning workers - consisting of unclothing and repeatedly washing, soaping and rinsing themselves off in the equipment bays. Despite these efforts, several firefighters had health problems in the months following, particularly skin problems and flu-like symptoms.

Again, in contrast to other areas, Chief Moreno reported that the GEMA flood maps were very helpful. However, it took some effort to secure them. While he commented that operations in general went much smoother than those after Hurricane Andrew, there was still a several day delay between ordering resources and receiving them.

In addition, local officials did not have a clear idea of their role at the EOC. Police officers, including Georgia State Troopers, were not clear about their responsibilities either, occasionally becoming simple bystanders to events. In Jones County, EMA Director Alan Greene had 100 fire department and 45 rescue squad personnel - all volunteer - available to help the 25,000 mostly rural residents of the county. Although just immediately north and east of Macon, and downstream of other problem areas, here the weather predictions were not receiving adequate attention.

By Tuesday afternoon, problems were starting to occur in the northern (more rural) part of the county. Three families were evacuated from their houses by boat. Personnel were also asked to assist a Monroe County search for two canoeists who were in turn looking for two men who had washed away in a car.

The Agency's "john" boat was quickly determined to be too small, though there were enough lifejackets available for personnel. Requests were made for assistance to the sheriffs office for patrol units, and to the Georgia State Troopers for a helicopter to patrol the river. One area of concern was some local earthen dams (now under survey by the U.S. Soil Conservation Service) and two Georgia Power dams on the Owugee River.

As in other areas of the state, there were communications problems. The previous April a statewide paging system was activated at a tower to the west. Pages constantly interfered with communications during the next 36 hours. Ham radio operators were able to help out substantially.

Fortunately there were no further water rescues. With only a 12-foot "john" boat and four lifejackets, rescuers were inadequately equipped. Further, standard GEMA rescue squad certification training covers water rescue only briefly, and no advance training was available.

A major effort was devoted to providing mutual aid to Macon County in the form of fresh water. Jones County provided nine trucks to transport water to Macon. The Georgia State Forestry Commission shops modified fittings on the water trucks to
allow them to deliver the water, While no fires occurred, some structures, including two mobile homes and one frame house, were lost to rising waters. The EOC in Jones County was not activated at all during the emergency.

II. LESSONS LEARNED

For the Incident Commander:

1. Establish regular training systems with local elected department heads, both in activation of the disaster plan, the correct functioning of the EOC, and the Incident Command System.

2. Much of what comprises the body of river and flood rescue technique is fairly specialized and unique. Get specific training for flood prone areas.

3. Pre-plan. Identify a local resource list of other trained agencies, location of additional equipment - particularly personal flotation devices for all rescue personnel - and boats and trained operators appropriate to swift-moving water.

4. Consider local enforcement powers, by ordinance, to cite drivers and homeowners who ignore warnings, evacuation orders, and road barricades.

5. Cooperate and train with state-level law enforcement authorities.

6. Utilize the National Weather Service. Err on the side of caution in ordering evacuations.

7. Fresh water for rescuers and residents will be a top priority. Have a resource list of water trucks and portable desalinization plants.

8. Floods should, be treated as a hazardous materials incident. Be prepared to decontaminate workers who must get into the water.

For the Emergency Operations Center:

1.   Make sure the dedicated EOC is large enough, on high ground, with good access, and that enough phone lines are dedicated.

2.   Conduct drills in the EOC.

3.   Design and use a 1/2 page phone report form for all incoming messages. Set up a message center in the EOC to pass critical information to any separate dispatch centers.

4.   Stock the EOC with flood plain maps and U.S. Geological Survey topographical maps. Street maps are not enough. Make sure personnel train in how to read a topographical map.

5.   Designate one ordering point in the EOC to avoid multiple ordering of the same materials, and failure to order other critical supplies.

6.   Route orders for materials delivery to a supply reception center, away from the EOC.

7.   Have stocks of emergency hand-held radios. Ordering such a resource from state authorities may take several days.

8.   Emergency managers need to remember that firefighters and rescuers - paid or volunteer - are a unique group in our society in their ability to overcome, adapt and improvise.

Appendix A

Safety Information for Georgia Flood Victims

Georgia Fire Academy

Georgia Public Safety Training Center
1000 Indian Springs Drive
Forsyth, GA 31029-9599
(912) 993-4670

FOR IMMEDIATE RELEASE: 7/11/94

SUBJECT: <u>**SAFETY INFORMATION FOR GEORGIA FLOOD VICTIMS**</u>

PREPARED FOR GEORGIA EMERGENCY MANAGEMENT AGENCY (per request of 7/10/94 P.M.) BY:

THE GEORGIA FIRE ACADEMY
GEORGIA PUBLIC SAFETY TRAINING CENTER
1000 Indian Springs Drive
Forsyth, GA 31029-9599

FOR FURTHER INFORMATION CONTACT: Don Ethridge
Supervisor of Fire Safety
Georgia Fire Academy
(912) 993-4670

STORY:
     Georgians are encouraged to consider the following precautions before moving back into flood-damaged homes:

-A qualified building inspector should first determine if the home or building is safe BEFORE you return to your property. A building inspector will check to see if there is structural damage. The house may have been moved by the flood waters. Shifting or settling may have damaged the structure or the foundation, as well as gas, water, sewer, and electrical lines. Houses that have been inspected and approved for repairs should be visibly marked on the outside by the building inspector.

-Exercise extreme caution OUTSIDE! Avoid downed electrical wires, leaning power poles, low-hanging wires ,or exposed underground cables. The electric utility should inspect all outside poles, wires, cables, on-ground transformers, electrical service entrance, and your electric meter and meter base. Notify utility companies before doing any digging. Georgia Power customers finding downed electric lines can report them by calling 1-800-353-1080.

-Electrical service to the house or building should be shut off by the utility if unsafe conditions exist on the property or inside the house.

FIRES COST LIVES AND MONEY

-Outside gas supply lines, including propane and butane tanks and service lines, should be checked for damage or leaks by the gas supplier or appropriate qualified personnel. Gas tanks actually float in flood waters, and service lines may be broken or damaged. Gas service should be shut off if leaks or damaged lines are found inside or outside.

-No motor vehicles, portable generators, electric or gas motors (such as for water pumps), cigarettes, welding torches, propane torches, lanterns, candles, portable stoves, barbecue grills, sparks, or fire/flames of any type should be permitted outside or inside the house or garage. Natural gas, propane gas, or butane gas from broken lines may be present. Methane gas from sewers/drains, gasoline vapors, or other invisible vapors may be present inside or outside. Fire or explosion could result. Have all areas checked first by qualified personnel.

-Expect the unexpected! In addition to wild animals, snakes, and frogs that may have taken refuge from flood waters in your house, you might also find gasoline cans, chemical containers, and propane gas cylinders that have been washed onto your property'. Gasoline produces invisible EXPLOSIVE vapors even on the coldest day. Gasoline floats on water and can penetrate wood and plaster/drywall. Expect to find it wherever flood waters went inside your home. It can easily travel from the lawnmower in the basement to an upstairs bedroom in a flooded home. Diesel fuel and other fuels and petroleum products float on water. Expect to find unexpected flammable liquids and vapors in your flooded home, especially in the basement.

-Gasoline is deadly. Even a small spark from a light switch or the pilot light in a hot water heater can ignite its vapors. Don't even consider using it for anything except fuel for gas engines. Mineral spirits is much safer for cleaning paint brushes. Gasoline should be stored only in locked, well-ventilated metal utility buildings outside the home.

-Dangerous chemicals may be present. Water can cause serious and deadly situations when it comes in contact with stored chemicals in and around the home. Linen closets, kitchen sink cabinets, basements, carports, and garages can be dangerous areas, and their contents can move to other rooms with flood waters. Chemicals often escape from their marked containers, giving no indication of their presence. Poisons, insecticides, herbicides, and cleaning compounds present obvious hazards. Drain cleaners, pool chemicals, bleach, lye, ammonia compounds, and other chemicals can produce skin burns and/or vapors that can kill or cause lung damage. The Georgia Poison Control Center can be reached by calling 616-9000 in the Metro Atlanta calling area or 1-800-282-5846,

-Water in contact with oxidizing compounds such as ammonium nitrate fertilizer can cause explosive reactions and/or fire. Bags of wet charcoal have reportedly caused fires due to spontaneous combustion. Remove them to dry and safe outdoor areas.

-All gas appliances that have come in contact with water need to be checked by qualified service personnel before being returned to service. Exhaust vents and chimneys must also be checked for obstructions, damage, rust, separations, and leaks (all the way to the roof). Defective or damaged exhausts/vents can result in deadly invisible carbon monoxide gas entering your home. Check with your gas supplier. In Georgia, Atlanta Gas Light Company, Georgia Natural Gas Company, and Savannah Gas Company provide free safety checks for their customers. Atlanta area customers can call (404) 522-1150. Savannah area customers can call (912) 944-2470. Customers in the Macon and Montezuma areas can call (912) 746-3090. Natural gas customers in Americus, Albany, Bainbridge, and other areas should contact their municipal gas suppliers.

-All electrical receptacles, junction boxes, switches, circuit breakers, lighting fixtures, and ALL other house wiring that has been exposed to water should be inspected by a state-licensed electrical contractor before power is turned on. Do not use ANY electrical appliances exposed to water until each one is inspected by qualified personnel. All heating and cooling equipment components must be checked. Failure to inspect and service electrical appliances after contact with water can result in serious damage to the equipment, as well as a very real chance of electrocution and/or fire. Get the wiring and the appliances inspected!

-Make sure that electric hot water heater thermostats are checked carefully by a qualified electrician. For both gas and electric hot water heaters, have the technician check for proper operation of the temperature and pressure relief (T&P) valve, and have the T&P drain line checked for obstructions, such as mud. T&P valves/drains, when properly maintained, reduce the chance of water heater rupture or explosion. They are important safety devices!

-Gas and electric stoves, as well as heating elements in clothes dryers, must be checked for debris that could start a fire. Clothes dryer exhaust vents should be checked for obstructions.

All space heaters exposed to water should be inspected carefully by a qualified technician. Do not use any type of space heater to attempt drying-out operations. They were not designed for this purpose. Fire or carbon monoxide poisoning might result. Fans and moving air are the most effective and safe way to dry out areas.

-Once the air in the house is determined to be safe, temporary lighting is best supplied by battery-operated flashlights. Candles, oil lamps, and kerosene lanterns require ventilation (a supply of fresh air). They can also be dangerous fire hazards!

-Camp stoves and gas lanterns should NOT be used indoors. Charcoal should NEVER be used for indoor cooking or heating due to the large quantity of deadly carbon monoxide that is produced, as well as the fire hazard of an outdoor grill used indoors.

-Before you start working INDOORS on repairs, make certain that you can get back OUTDOORS in a hurry if problems develop! Draw a new Home Escape Plan (unless your old one survived the flood) and practice it before you start repairs. Check EVERY door and window to make certain that it will still open easily if you have to use it as an emergency exit. Flooding will often cause wood to swell and other stresses that will cause doors and windows to bind. Make sure that they all work. Have a meeting place designated at a safe location in front of your home. Fires happen in seconds. GET OUT and STAY OUT. Call 911 or other emergency numbers from your neighbor's house, or use a cellular phone.

-If you suspect that water or moisture has come in contact with your smoke detectors, replace them immediately! Battery-operated smoke detectors now sell for less than $10.00!!! Buy several and install them immediately, following the manufacturer's instructions. Do not spend a single night in your home without several working smoke detectors. Vacuum the dust from your smoke detectors and test them daily while construction and repairs are in progress. Your smoke detector can't take care of you if you don't take care of it. Keep extra batteries on hand for your smoke detectors. Batteries normally last about one year. You will need to change them before the anniversary of this flood!

-Use extreme caution when using portable generators to supply temporary electrical power. Portable generators should be located safely away from the house, and they should be properly grounded. Read the instructions and get qualified advice. Never re-fuel any gasoline-powered engine while it is hot. Do not attempt to connect portable generators to the house electrical system. This can endanger you, your family, your house, and power company employees working on lines miles away from your house!!! Don't do it! Remember that portable generators produce enough voltage and current to kill you. Exercise good electrical safety practices.

-During repair operations, use only double-insulated or properly-grounded electric power tools. Extension cords should be rated higher than the current draw expected. Only grounded cords should be used. Discard damaged cords. The use of fixed power outlets equipped with ground fault interrupters (GFI'S) or portable GFI devices while using power hand tools, vacuums, etc., will provide an added margin of safety, indoors or out.

-Be especially careful when working or staying in wet or damp areas. Your body can become an electrical conductor if you touch any source of electricity, even a defective lamp or appliance.

-Last but not least, remember that your damaged home is probably NOT a safe location for your children just yet. If children do come on the site, make absolutely certain that they are supervised at all times by an adult who is not engaged in the repair and clean-up operation. Keep children out of work areas. Wait until repairs are complete and your home is safe before moving in the family. Be prepared to start your emergency plan if it is needed. After all, you are recovering from one disaster now! Try to avoid another due to fire or avoidable injuries.

*U.S. GOVERNMENT PRINTING OFFICE 1996 - 719-595/82752